Pezibear "Pixabay"

I dedicate this book to my mother,

who passed away while I was writing it,

and to my beloved daughter.

My darling, you are my inspiration,

and I have always loved you dearly.

Mum

WHY CHOOSE HOMESCHOOLING?

Why Choose Homeschooling?

ANGEL CLAUDIUS

c.i.t.e

Copyright © 2025 by Angel Claudius, Australia
Claudius Institute of Training and Education Imprint
www.angelclaudius.com; www.citepublisher.com
Acknowledgments: All images have been obtained from "Pixabay" and are used under the terms of the "Pixabay" Licence Agreement
A catalogue record for this book is available from the National Library of Australia, ISBN 978-0-6458672-4-4
All rights reserved. The right of Angel Claudius to be identified as the author of this work has been asserted by the Copyright Act 1968 (Australia). No part of this book may be reproduced, stored in a retrieval system or transmitted in any form or by any means without the written permission of the author and publisher.
First Printing, 2025
Limit of Liability / Disclaimer of Warranty:
The publisher and the author do not guarantee the accuracy or completeness of the contents of this work and specifically disclaim all warranties, including, but not limited to, those regarding the subject matter covered. No guarantee can be created or extended by sales or promotional materials. The advice and strategies provided in this work may not be suitable for every situation. This work is provided in good faith with the understanding that the author and publisher are not providing educational, medical, legal, or other professional advice or services. Neither the publisher nor the author shall be liable for any damages arising from this work. The inclusion of an individual, organisation, or website as a citation or potential source of further information does not imply endorsement by the author or the publisher of the information or recommendations provided by that individual, organisation, or website. Readers should be aware that websites listed in this work may have changed or become unavailable since this publication was written and published online or in print.
A note on gender identities:
The author and publisher recognise that individual experiences and interpretations of gender can vary widely. In the context of gender, language and its application in society, the scientific and medical communities are continuously reevaluating their terminology. Most of the studies referenced in this book use "women" and "girls" to refer to individuals assigned female at birth, whereas "men" and "boys" refer to individuals assigned male at birth. In this publication, the term "child" or "children" may refer to either group. To the fullest extent permitted by law, the author and publisher disclaim any liability for the use of the information presented in this book, including those related to gender, as such topics are sensitive and subject to personal choice.
Photo credits: "Pixabay" Frt Cover-TyliJura; Bk Cover-Pezibear

CONTENTS

Dedication i
Why Chose Homeschooling? viii

1 **Introduction** 1

2 **Why choose to homeschool?** 10

3 **Why choose to homeschool? - Media Resources** 21

4 **Find inspiration and motivation to homeschool** 33

5 **Conclusion** 41

6 **Resources: Useful Sites** 46

Homeschooling publications 50
References 53

Why Chose Homeschooling?

TyliJura "Pixabay"

"Valid Reasons for Home Education Supported by Media Stories"

"If we let our children to cross the irreversible red line of mainstream schooling's negative impact in the name of education, the damage done to them would be impossible to undo."

By Angel Claudius

| 1 |

Introduction

u_r6sfyh2sbr "Pixabay"

"I am truly excited to help others on their homeschool journey, as I have walked this path myself. You are now welcome to learn from my experience".
(By Angel Claudius)

H ello, my name is Angel, and I'm an author, designer and home educator. I wrote this book to inform, encourage and motivate parents and caregivers who are considering homeschooling* or are already doing so. My first encounter with homeschooling took place many years before I had a child of my own. A friend of mine, who homeschooled her four children, shared some valuable advice with me. She said that from a child's earliest years, parents have the opportunity to significantly influence their life and shape their character through home-based education. If parents become actively involved in their children's lives from that time, they can help them become well-balanced individuals rather than leaving that responsibility to strangers. This advice made me think and resonated deeply with me, and I kept it in my heart for the day when I would become a parent.

Years later, I overheard two primary school girls in the swimming pool's changing room discussing their desire to start homeschooling. One girl expressed that she did not want to go through the stress of waking up early every day for school and being bullied there. She also loved the idea of being able to stay in her pyjamas for as long as she wanted and feeling relaxed. As they discussed their negative experiences with school and how they had impacted them, I couldn't help but reflect on these conversations. At the time, my child was in her Prep year at a private school, and I was seriously considering homeschooling as a solution to the challenges we were facing there. I have been homeschooling my twelve-year-old child since kindergarten, except for a brief period when she attended a private Montessori school for about six months. Although I initially believed that a private school would provide her with a superior education, I eventually realised this was not the case for us. After struggling for various reasons, my daughter's dissatisfaction was very obvious, and her pleas moved me to take her education back home. While in private school, she developed anxiety attacks and chronic stomach pains, which disappeared after several months of treatment. When she returned to her home environment, she gradually began to feel like herself again. As a result of our negative experience with school, we decided to continue our child's education at home.

I love the topic of home education because I've seen many benefits of this type of schooling in my own child's life. I came to realise that the beauty of homeschooling lies in its flexibility and numerous advantages, including the ability to influence and tailor the child's curriculum. When considering homeschooling for our children, we must account for the associated costs and determine whether it aligns with our family's priorities and circumstances. Parents should also reflect on their values when deciding whether homeschooling is the right choice for their family's situation. This decision is very personal and depends on several factors, including the child's personality. As a family that opted for homeschooling, we have had to make both personal and financial sacrifices to make this choice a reality.

During my homeschooling experience, I always believed the motto "*Where there is a will, there is a way.*" We overcame many obstacles with hard work and established a practical, flexible daily routine for our family. Furthermore, I compiled a list of reasons for choosing to homeschool our child. This list has proved extremely useful in motivating and inspiring us during challenging times of homeschooling. Yes, we must be aware of our reasons for home educating our children and remain confident in our decisions at all times.

Some people think they cannot give up their professional careers to homeschool their child. But it is unfair to assume that stay-at-home parents and home educators have less valuable work than those with paid jobs. Being a stay-at-home parent and homeschooling educator indeed needs significant attention and effort, just as any other professional career, but in return, it brings priceless benefits! These homeschool responsibilities have no set start or end time, and we must realise how serious and essential these duties are. Some parents may also hesitate to homeschool because they feel unqualified as teachers. Although most home educators do not have formal teaching qualifications and face other life challenges, they are committed to providing their children with a personalised, safe, and enriching education at home. In essence, the key to successful home education is having loving, willing parents (or a parent) who prioritise their children's self-paced, flexible and happy education at home. As Maria Montessori noted, (Quote) "*home education is the work of love*" (end of quote). In addition, I can't emphasise enough how important it is for parents to be willing to learn alongside their children and to remain open-minded and adaptable.

During our home education journey, I learned to empathise with my child by *putting myself in her shoes* and trying to understand her better. But it does not happen automatically and requires effort. The homeschooling experience revolutionised my approach to parenting, leading me to transition from a fashion designer to an author and book designer. Devoting more time to my child allowed me to understand her needs and adjust homeschooling methods accordingly to fit her personality and our family's lifestyle. I have also observed improvements in my cognitive abilities since I began learning alongside my child. But starting something new in an unfamiliar field, such as homeschooling, can be daunting! So, the key is to remain motivated and committed, and never give up on your decision once you've made it. Although family situations may change, you can continually reassess and make the best choice for your child (or children) and family.

For a dedicated home educator, it is essential to be familiar with the homeschooling regulations in your country. Acquire effective teaching methods to enrich your child's learning experiences and support their growth into mature individuals. Take a few minutes to learn about homeschooling teaching methods in the quiet of the evening. It will help your child become happier, more responsive, and fully benefit from your efforts. If your child is happy, it is a sign that your homeschooling efforts are successful! As Maria Montessori once said, (Quote) *"One test of educational procedure is the happiness of the child"* (end of quote). Remember, to make every moment count.

Make an effort to review your homeschool routine, setup, time management, and teaching techniques regularly. If you identify areas that need improvement, take immediate action to fix them. With time and practice, you'll become a more confident home educator. Your child will value their traditional, family-oriented upbringing both now and in the years to come. While homeschooling, children can develop their skills and desired personalities in a relaxed and familiar environment. They will have more time to play and enjoy their childhood. Most importantly, homeschooling can strengthen the emotional bond between you and your child. What more can you ask for?

One of the main benefits of homeschooling is the ability to personally teach and instil your family's values more fully in your child's life. You could even incorporate these values into your homeschool curriculum. However, you may feel mentally and even physically drained during your initial attempts to assist your child with teaching. This is likely to happen, especially if using traditional, old-school methods. In that case, review your homeschool rhythm. Aim to create a natural flow of your homeschool time that helps establish a balance between structure and flexibility, as well as between intentional learning and natural discovery in your home-based school. Additionally, staying motivated and reminding yourself of why you chose this path will be incredibly helpful during the challenging times you are likely to face.

Homeschooling is not just about acquiring and transferring knowledge and skills to your child; it is also about nurturing relationships with others along the way. In many developed countries with high standards of living, such as Australia, it is common for children to leave their loved ones and parents' home at a young age before reaching maturity and becoming independent. It can be distressing for parents to see that happen, and in some instances, even disadvantageous to the children. I want to ensure that this scenario does not happen in our family. As an old proverb says, "*prevention is better than cure*". By strengthening the bond and connection with our children while they are in our care, they are more likely to benefit in the long term by staying with their parents for an extended period.

Consequently, homeschooling can provide a strong foundation for a fulfilling life.

Home education faces challenges in various parts of the world and has its drawbacks. This book examines the key reasons to homeschool, drawing on recent media coverage to support its arguments. It will also help you assess your circumstances to determine whether this type of education is suitable for your family. In every endeavour we pursue, there are always two sides to the story. It is the same with home education, according to "Teen says homeschooling helps her learn, but academics warn of pitfalls" ("ABC News Australia", 19 Sept 2025)[1.1]. A teen has tried four schools, but each time she left, as she knew that (Quote) "it wasn't the right fit". She said (Quote) "I didn't like traditional schooling, mainly because you're kind of just put on the spot, and you're expected to do all these things" (end of quote). She said she has benefited from homeschooling. (Quote) "I wake up and I'm not absolutely dreading going to school. I'm not dawdling in the morning" (end of quote). Professor Anna Sullivan said that homeschooling is not for every child. (Quote) "Once a family commits to homeschooling, sometimes it can be tricky to transition children back into the mainstream (...) Education is not just about learning content and has a bigger purpose (...) Homeschooling can be really beneficial for children who get overstimulated by large groups" (end of quote). But the teen said that it made her more confident. (Quote) "It's not just about doing maths or literature. It's also being out and learning life skills (...) it's about just getting out and doing what you love" (end of quote).

Tarasyasinski "Pixabay"

Recently, I read a single parent's expressions of the damaging impact that a traditional, mainstream school environment has had on her children in a story, "I homeschool my kids-parents like me are judged and exhausted." (Quote) "I have witnessed one daughter pulling out her eyelashes and rocking in a corner, another curled in the footwell of the car, trembling and mute. I will never unsee those memories" (end of quote) ("MSN", 1 Sept 2025)[1,2]. She said her children have been sadly traumatised within the school system, and concluded that a traditional school setting, however nurturing the school, doesn't suit all children. Homeschooling was a necessity for her, and you may feel a similar way. Whatever the reason, if you're reading this book, you've likely decided to try homeschooling. You are now committed to meeting your child's unique needs. Although working as a home educator may be challenging at first and raise concerns, you will ultimately be comfortable with your decision. Always remember the bigger picture, remember WHY you are doing this, stay motivated, and never give up, regardless of the obstacles you face. If you experience challenges with homeschooling, find practical solutions and implement them promptly in your current situation. However, determining what type of education is most effective for your child and family is ultimately up to you. This publication serves as an information guide only, and conducting thorough research at any stage of the homeschooling process is crucial for all readers considering home education. However, if you are willing to give it a try, congratulations on making this brave and proactive decision to homeschool, and thank you for using this publication as a guide at the start of your journey.

Tarasyasinski "Pixabay"

*(In this book, the terms *home education* and *homeschooling* refer to the education of a child at home by at least one parent or caregiver who follows local laws and regulations).

| 2 |

Why choose to homeschool?

"Never let your traditional formal education get in the way of your child's opportunity to natural learning" (By Angel Claudius)

Tarasyasinski "Pixabay"

A child can be likened to a tree. It can grow tall and strong, provided its roots are deep and healthy, and it receives regular water and nutrient supply in its early stages. Similarly, home education can play a vital role in a child's successful growth, development and learning experiences as well as in shaping the whole person. Home education provides vital 'nourishment' through many benefits, including increased family time, the opportunity to instil the family's moral and spiritual values, and greater community involvement. Parents who commit to homeschooling their children do so out of sincere concern for their children's well-being and out of love. They carefully customise an educational plan for each child, taking into account their unique personality, individual needs, and family circumstances. Doing so enables each child to reach their fullest potential. They also help their students by emphasising their strengths and offering additional support in other learning areas as needed. Above all, homeschooling allows each child to learn at their own pace, resulting in a truly personalised educational experience.

As a home educator, you have the freedom to incorporate a variety of activities into your child's education. These include tutoring, extracurricular activities, online courses, field trips, religious education, sports, nature studies, work experience, volunteer work, various events and others. Teaching your children at home provides you with greater flexibility to tailor their education to their individual needs and interests. What's more, homeschooling can strengthen family bonds. Ultimately, homeschooling encourages children to think independently, act for themselves, pursue their own goals, and maintain mental and emotional balance. All these factors help shape a child into a mature individual. What a priceless benefit and reward!

Interestingly, there was no public education system in the past, and children were taught at home by their parents. Sons were taught by their fathers to earn a living, while daughters learned from their mothers many practical skills and how to care for their households. In some, girls were also taught to play musical instruments. These children could learn naturally and use their skills to survive in society. Most children today receive their education in traditional mainstream schools. However, the way they continue to learn naturally through experiences remains essentially unchanged.

Some parents view home education as an opportunity to instil the family's values in their children without distractions and external pressures, while others are motivated by religious or moral beliefs. Yet, sometimes, homeschooling might be the only option due to the child's specific needs, negative social influences such as bullying, or financial factors. Regardless of the reason, homeschooling demands a significant commitment.

Successful homeschooling requires four essential elements, which are the main ingredients of a successful home education recipe:

1. Parents' solid desire to commit to teaching = **M**otivation

2. Children's willingness to accept parents as their primary all-day teachers = **A**cceptance

3. Parents' willingness to learn alongside their children, and children's willingness to study and learn at home = **A**ttitude

4. The family's favourable circumstances = **S**ituation

5. A good* parent-child relationship = **C**ooperation
(*or reasonably good)

= **MAASC**.

> "The function of education is not to give technical skill but to develop a person; the more of a person, the better the work of whatever kind". (Quote by Charlotte Mason, "Towards a Philosophy of Education, Vol . 6, p.147)[2.6].

One of the most appreciated benefits of homeschooling is that parents have the freedom to instil in their children the family's good moral values and respect for others. At the same time, they can ensure their children receive a proper, good-quality education. By receiving personalised attention, children can focus on areas where they require additional support, thereby reducing the risk of falling behind. This helps them learn at their own pace and avoid the pressure of being compared with other students in the class. Homeschooled children also receive education through interactive experiences with peers of different ages, adults in their community, parents, siblings, and online or in-person tutors. Additionally, homeschooling significantly reduces both physical and mental stress, enabling children to thrive in all aspects of their lives and maintain better mental and physical well-being. This is in strong contrast to the children attending traditional mainstream schools. The majority of these children experience high levels of anxiety and stress. Combined with their worries, this could potentially lead to long-term mental health issues for them. These children must face overwhelming social anxieties, which cause some to fall behind academically. But, home-based education involves a lifestyle in which academics are only a minor part of the day, leaving sufficient time for natural, relaxed learning. Homeschooled children are recognised for being self-motivated, happier, and excelling in all aspects of life. They also develop balanced personalities and are generally better equipped to handle life's challenges. Homeschooled children have ample time to play, explore their interests, and engage in enjoyable activities. As a result, they tend to experience greater life satisfaction in childhood and adulthood. This contrasts with children in conventional schools, who often endure physical and mental stress because of the pressures of traditional academics and problems in personal life, resulting primarily from association with peers of the same school. Additionally, some may face bullying, while others feel compelled to conform to beliefs they or their families do not agree with or even actively oppose. A common outcome for these students is the paralysis caused by social anxiety.

Pezibear "Pixabay"

Moreover, students attending traditional mainstream schools often face overcrowded classrooms with thirty or more students and other unreasonable restrictions. I discovered that in a particular primary school in Melbourne, two classes of approximately 30 students each are separated by a divider that is removed each school day to form a combined classroom with only one teacher! How could he possibly give any one of these students in this large classroom his personal attention?

In addition, there are increasing dangers in public schools, one of which is school violence. Years ago, I met an English teacher who worked with elementary-level students at a mainstream primary school. She said that her last day of work there was the one when her student took a knife out of his pocket to threaten her life! Unfortunately, in some countries, parents are fearful of sending their children to school, and teachers feel apprehension when dealing with their students in the classroom. In addition, apart from mental and physical stress, other dangers present themselves. An example of this is bullying, school shootings, vaping, drugs, gender issues, dating and teen sex before reaching maturity, identifying as animals and others. These concerning issues seem to be on the rise globally.

All to-be home educators must also be aware that homeschooling is a lifestyle that can significantly impact their daily routines and relationships. For example, you won't have as much time for yourself as you used to, your house may be constantly messy or noisy, etc. Despite that, homeschooling provides unique and priceless benefits, such as a customised education tailored to each child's needs, regardless of age or grade level. This is one of the most significant advantages of homeschooling and one that many families find priceless.

> "How do you undo the damage caused by what the child experiences at school?"

In addition, home education presents numerous other benefits, such as:

- Excellent self-paced education.
- Dedicated, virtuous parents who are intimately involved in their children's education.
- Effective teacher-to-student ratio.
- Education tailored to a child's capabilities, special needs, interests and personality.
- Consistent and integrated education in a family. This involves lessons covering multiple subjects or learning areas, suited to students of different levels.
- Engagement in a family-oriented lifestyle, which enables homeschooled children to grow up with a traditional, old-fashioned, happy childhood.
- Use of effective teaching methods and materials suited to the child's individual needs and capabilities.
- Facilitating young people with engaging cultural experiences, like family traditions, customs, as well as lessons on moral and spiritual values, to a fuller scope.
- Constant parental supervision and attention.
- Direct parental influence in all aspects of children's lives.
- Ensuring children's personal safety, balanced emotional, mental, and physical health and more.

Sadly, the mainstream education system, whether public or private, cannot possibly meet these benefits. On the contrary, an unsafe environment rife with problems negatively affects children's overall well-being, which is another factor in favour of home education. I recently read a book by Ann Block titled "*No More ADHD*". According to the book, many parents were pressured to medicate their children with potent, medicinal drugs to treat ADHD-like symptoms, which come with a long list of unwanted side effects, to ensure they sit quietly in the classrooms for their teachers' convenience. The story "Millions of kids are on ADHD pills. For many, it starts a drug cascade" ("The Wall Street Journal", Apple News +, 27 Nov 2025), shows that (Quote) "Powerful drugs are often the next step" (end of quote). But, according to the story, "Experts reveal side-effects of prescribing ADHD medication too early in a child's life" ("The Independent", 1 Sept 2025)[2.1], (Quote) "Experts warn that drugs for attention deficit hyperactivity disorder (ADHD) are being prescribed too quickly to preschool children in the USA. (...) Early medication for young children can lead to more side effects and a higher likelihood of treatment failure, as their bodies process drugs differently" (end of quote). Dr Block highly recommends homeschooling in her book.

The homeschooling movement is gaining popularity as it returns the responsibility of education to parents, who were initially entrusted with this role in society. As more families become aware of the benefits of home education and seek the best possible learning experiences for their children, the number of homeschooling families continues to increase steadily. A friend who had worked as a primary school teacher for most of her professional life once told me, "*How do you undo the damage caused by what the child experiences at school?*" It's a sobering question every parent should ask themselves. Yes, homeschooling is the way out of the mainstream education system's labyrinth.

There are a variety of homeschooling approaches, which are discussed in Chapter 5, "Know different homeschooling approaches" of the publication titled: *"Homeschooling in a nutshell: Effective advice for busy parents to achieve success in home education"*, ISBN 978-0-6455541-1-3*. A few homeschooling approaches include *unschooling*. According to "Unschooling movement has parents teaching kids outside of classrooms" ("NBC News", 21 August 2025)[2.2], a growing number of families are *unschooling* their children. Self-directed education is based on activities and life experiences chosen by the child, with the guidance of parents, within their boundaries. (Quote) "Unschooling is trusting yourself and trusting your child" (end of quote) said the interviewed parent. It has no lesson plans or testing; instead, it is a self-directed education approach that allows children to learn while following their own interests in their everyday environment.

I want to share the experience of the 68-year-old novelist Anthony Horowitz, who revealed his personal feelings about his education in a private school, which caused him (Quote) "incalculable damage", as presented in the "BBC" podcast. The author of the "*James Bond*" and "*Alex Rider*" series explained that he was sent to a boarding school at the age of eight. He mentioned that upon returning many years later, he felt like (Quote) "a terribly damaged person". He suffered from beatings that almost caused him to pass out. He claimed that parents are (Quote) "destroying their children" by sending them to private schools. In an interview with the "BBC" Headliners podcast, he revealed his experience ("Daily Mail", 18 April 2024)[2,3]. (Quote) "I was damaged. I mean, my schooling, the five years I spent at a school in North L......, did incalculable damage (...) And it is one of the weird things (...) that parents should pay so much money and put so much effort into destroying their children" (end of quote).

On a positive note, I would also like to highlight the story by "Business Insider" (17 May 2025)[2,4] titled: "Our friends and family discouraged us from homeschooling our kids. Now grown, one of them is a pilot and the other went to college on a full scholarship". In this article, we read (Quote) "Early on, my husband and I thought homeschooling might be the best option for our two sons. (...) Our family and friends discouraged us, but we decided to move forward despite their warnings. (...) We had our challenges, but both of our kids thrived. Today, they are both successful young adults" (end of quote). In another story by "Sydney Morning Herald" (20 July 2024)[2,5] titled: "I can't learn: why some parents have swapped school for homeschooling", we read (Quote) "It took a mother of three until her eldest reached year 11 to realise that mainstream school was doing her children more harm than good" (end of quote). So, she turned to homeschooling as a solution. Each child struggled with the teaching style at school, so the mum decided to become both mother and teacher. She said (Quote) "It was the best thing we ever did (...) For their mental health, us as a family and their understanding of who they are and how they learn" (end of quote).

According to *Queensland University of Technology Education* researcher Dr Rebecca English, this case is one of a growing number of "accidental homeschoolers," who account for about 85% of the sector. She said (Quote) "These are families who never intended to homeschool, but for reasons such as school refusal, neurodivergence, bullying, or just having kids who are different, prompted parents to look for alternatives" (end of quote). In the statement by K. James from the "Homeschool Education Network", we read (Quote) "Some children with sensory issues can't deal with noise or uniforms that are uncomfortable or scratchy, or they struggle with bright lighting (...) When a child is in their home, they are in an environment that is comfortable for them" (end of quote).

In Chapter 3 of this book, I have presented media stories that highlight the following problems and challenges (among others) of educating children in the mainstream education system: child sexual abuse, rising school costs, gender issues, teacher shortages, ideology taught at schools, broken school system, violence and aggression, excessive worries, anxiety and stress experienced by children, bullying, children with special needs suffering, pressure to date and have sex in teen years, dangers while travelling to and from schools, conditions at schools, school children identifying as animals, and others. As you continue reading, you may discover your reason—or even a few reasons — in favour of homeschooling in your family.

To learn more about the brief history of education, visit: https://www.familyeducation.com/school-learning/home-schooling-isn't-new-brief-history-education.

*The publication ISBN 978-0-6455541-1-3 is available from https://www.angelclaudius.com; https://www.amazon.com

| 3 |

Why choose to homeschool? - Media Resources

"Why send the kids to school when you can homeschool?" (By Angel Claudius)

Tarasyasinski "Pixabay"

Home education has a solid foundation: a well-informed decision made to meet the child's educational, emotional and physical needs. However, when considering the typical mainstream school environment, we must honestly acknowledge that children exposed to it often face various challenges and dangers, particularly in certain countries. While there is a wealth of information on traditional school topics, I want to highlight several newspaper articles I recently came across that present strong arguments in favour of homeschooling for our family, and you may feel similarly. (Please read *Disclaimer*).

1) Child sexual abuse

1.1) *"Fear, shame, anger: child abuse inquiry hears of agony"* ("Daily Mail", by William Ton, Oct 2023)[1.1] story is about child sexual and physical abuse by staff who worked at twenty-four Victorian schools. (Quote) "Fear, powerlessness, shame and anger. These are the lasting impacts of victims of child sexual abuse continue to experience after allegedly being abused by staff who worked at twenty-four Victorian schools".

2) Rising school costs

2.1) *"Growing number of australian families struggling to afford their kids' education"* ("ABC News", 13 Jan 2025)[2.1], (Quote) "The cost of educating children is a major financial burden on families-even at public schools-and the expense is much higher than just a year ago for families in major cities (...)".

2.2) *"Public school parents: are you aware of the underfunding of public schools? (...)"* ("Reddit", 30 Apr 2025)[2.2].

2.3) *"Cost of living squeeze hits school families"* ("The Educator Australia", 19 Mar 2025)[2.3]. "Independent school families are facing mounting financial pressure with many making significant sacrifices to keep their children enrolled (...)".

3) Gender issues *(Please read Disclaimer*)*

3.1) "*School children asked to complete 'outrageous' survey on their gender and pronouns*", ("The Australian", 29 May 2023)[3.1], a story by Natasha Robinson was broadcast on "The Sky News Australia" (30 May 2023). (Quote) "Public schools have become dangerous places" for children in Australia because (at the time this article was written), they help children transition their gender without their parents' knowledge.

3.2) "*Clutching at any diagnosis: Parents look for help with children's mental health*" ("Sky News Australia", 5 Sept 2023)[3.2]. (Quote) "PR Counsel Managing Director Kristy McSweeney says parents are 'clutching at any diagnosis' for their children experiencing gender dysphoria after seeing social media posts where people are ending their own lives over underlying mental health issues".

3.3) "*Scary stuff: Hospitals send 'gender-confused' kids to undergo transition*", ("Sky News Australia", 5 Sept 2023)[3.3]. Host Rita Panahi said (Quote) "Gender-confused children being sent to gender clinics to undergo irreversible reassignment surgery is 'scary stuff' (...)."

3.4) "*Virtue signalling: High School converts female-only toilet block to unisex*" ("Sky News Australia", 21 Apr 2024)[3.4]; Parents have criticised the lack of consultation over the move and are claiming their daughters are too scared to use the bathrooms.

4) Teacher shortages

4.1) "*Victorian schools facing mass teacher shortages*" ("Sky News Australia", 21 Mar 2024)[4.1].

4.2) "*Australia in the middle of an education staffing crisis*" ("Sky News Australia", 9 Aug 2023)[4.2]; It highlights the dilemma schools face due to a teacher shortage. (Quote) "So many schools are going without the teachers they need to cover the classes they have for their students", which leads to overcrowded classrooms and a lowered quality of education.

4.3) "*Thousands of Queensland teachers have gone on strike with 600,000 school students affected. What happens next?*" ("The Guardian", 6 Aug 2025)[4.3]. (Quote) "(...) An estimated 600,000 students in state schools were affected".

5) Ideology taught at schools

5.1) "*Walking billboard: Victorian School's new uniform has a political agenda*" ("Sky News Australia", 10 July 2023)[5.1].

5.2) "*Parents are 'desperately' concerned about ideology taught in schools*" ("Sky News Australia", 13 July 2023)[5.2]; Former Deputy Prime Minister John Anderson discusses this topic.

6) The school system is broken

6.1) "*Shouldn't be surprising: The School system is broken*" ("Sky News Australia", 9 Jan 2023)[6.1]; (Quote) "The school system is 'broken' because students are not receiving the education they need and are acting out due to a lack of direction (...)."

6.2) "*One in three not meeting numeracy and literacy expectations*" ("ABC News" (AU), 23 Aug 2023)[6.2]. (Quote) "One in three school students is failing to meet minimum literacy and numeracy standards across Australia, and one in ten are so far behind they need additional help".

6.3) "*Misery taught in schools stops students from 'striving to achieve'*" ("Sky News Australia", 23 Aug 2023)[6.3]. (Quote) "One in three students not meeting minimum expectations".

7) Violence and Aggression

7.1) *"Concern after ex-students enter school with weapons"* ("The Sydney Morning Herald", 12 Aug 2023)[7.1]. (Quote) "Concerned parents contemplate pulling their children out of a school in Melbourne's North after intruders stormed the campus, wielding weapons (...). It's the latest in a spate of youth-related violence across Melbourne. The number of children committing crimes almost doubled in the past year".

8) Anxiety, Worries, Stress

8.1) *"20% of primary school children suffer high anxiety"* ("Daily Motion", 17 Aug 2023)[8.1]. (Quote) "New data suggests there's been a dramatic rise in the number of primary school children suffering from high level of anxiety, which could be affecting as many as 400,000 families across the country. Research (...) shows that almost three-quarters of teachers believe they don't have the skills to help".

8.2) *"Children's worries could turn into long-term mental health conditions if not managed"* ("Sky News Australia", 4 Sept 2023)[8.2]. (Quote) "Kids' worries turn into long-term mental health conditions if they don't have an opportunity to talk to a close adult".

8.3) *"Growing number of school students becoming paralysed by social anxiety"* ("Sky News Australia", 28 Nov 2023)[8.3]. (Quote) "Some kids are skipping school and falling further behind academically. Even high-achieving students are finding themselves unable to attend regular classes due to anxiety (...) They've got all sorts of social anxieties. Some are falling behind so far that they don't want to go to school anymore".

8.4) In a story by "Mouths of Mums"*, (written by Anita Butterworth, 10 May 2024)[8.4], a member admitted she didn't know what to do after her 12-year-old son ended up wetting himself when his teacher denied him a toilet break.

9) Bullying / AI bullying

9.1) *"Complaints of children using explicit AI images to bully peers"* ("Dailymotion", 20 Aug 2023)[9.1]. (Quote) "The eSafety commissioner's office has received its first complaints of children using AI to bully their peers".

9.2) *"Aussie mum makes global plea at UN to kick off social media"* ("Dailymotion", Sept 2025)[9.2]. An Australian mother appeared at a United Nations event in New York after her teenage daughter died by suicide after online bullying.

10) Children with special needs suffer

10.1) *"New Data shows disabled children experience high rates of bullying, exclusion, and lack of support"* ("ABC News", 20 Aug 2023)[10.1]. What is it like for students with disability in school settings? (Quote) "They experience high rates of bullying, exclusion, and a lack of support in classrooms; 70% of students with a disability felt excluded from activities and events at school; 65% reported bullying at school, (...) Parents are desperate (...)".

10.2) *"Parents moving bullied children to other schools as teachers struggle to control student behaviour"* ("ABC News", 20 Nov 2024)[10.2].

" Misery taught in schools stops students from 'striving to achieve' " (*Sky News Australia*)

11) Dangers while travelling to and from schools / During excursions

11.1) *"School bus rolls in Victorian town"* (<u>"SBS"</u>, 27 Aug 2025)[11.1]. (Quote) "(...) 1 student dead, 24 injured, including serious injuries like fractured skull and broken bones".

11.2) *"Croydon: Girl, 15, stabbed to death on way to school - as teenage boy is arrested."* (<u>"Sky News"</u>, 28 Sept 2023)[11.2].

11.3) *"Six students drown during school trip to Egyptian beach"* (<u>"Daily Mail"</u>, 24 Aug 2025)[11.3].

11.4) *"Student who died in Bendigo after being hit by a truck remembered as being 'smart and dedicated'"* (<u>"ABC News Australia"</u>, 12 Oct 2025)[11.4]. A Bendigo student was killed on his way to school when he was hit by a truck.

11.5) *"Two children were struck by cars near R....Hill Public School before death of Islah Metcalfe"* (<u>"MSN"</u>, 14/11/2025)[11.5]

12) Conditions at schools / Safety

12.1) *"Protest erupts over school's prison 'rules' as students suspended for being two minutes late and asking to use toilet"* (<u>"LBC"</u>, 21 Sept 2023)[12.1]. (Quote) "Parents said the strict rules have seen children put in *reflection* for minor infractions, including yawning, going to the toilet, and having laddered tights (...)".

12.2) *"Twin tragedy: Dad's grief as 4-year-old is crushed to death at school birthday party"* (<u>"The Sun"</u> / <u>News.com.au,</u> 11 Aug 2025)[12.2]. A heartbreaking tragedy unfolded when a four-year-old girl was fatally crushed in a school playroom accident, with her twin brother witnessing the incident. The father said (Quote) "(...) We lost our daughter in the place where she should have been safe and protected".

12.3) "*Parents of Melbourne schoolboy Jack Davey tell court their son's life was 'stolen' by driver*" ("ABC News Australia", 1 Sept 2025)[12.3]. (Quote) "Jack, 11, was killed and four others were injured when an SUV crashed through a fence and struck a table at A..... S.... Primary School (...) Jack was crushed between the table and the SUV, suffering fatal head and chest injuries".

12.4) "*Parents' anger at new 'prison-like' rules at school*" ("Metro", 9 Sept 2025)[12.4]. (Quote) "Parents have complained of a new 'prison-like regime' at a secondary school, where pupils have ten minutes to eat their lunch and toilets are locked during lesson time."

13) School refusals

13.1) "*School refusal and the Australian families gripped by blame, shame, and the fear of a lost education*", "*The kids who fear school*" ("ABC News" (AU), 29 Apr 2024)[13.1]. (Quote) "A growing crisis of school refusal is gripping Australia (...). Since school attendance is mandatory by law, parents of children facing challenges in going to school are always under the looming threat of legal action".

13.2) "*School refusal closely related to anxiety in young children and pre-teens*" ("SkyNews.com.au", 22 Mar 2023)[13.2].

14) School shootings

14.1) "*Almost 1,700 kids and teens were killed by guns in the US in 2023*" ("The Daily Digest", 12 Mar 2024)[14.1]. (Quote) "Over 80% of school shooters stole guns from family members. Last year (2023), a six-year-old student brought a gun into his classroom and, without hesitation, shot and wounded his teacher."

14.2) "*Former student identified as gunman who killed 10 in school shooting in Austria*" ("Euronews", 10 June 2025)[14.2].

(Quote) "Ten people were killed in a school shooting in the Austrian city of Graz. (...) At least thirty people were admitted to the hospital for treatment (...) with some seriously injured".

14.3) *"Two children killed in Minneapolis school shooting"* ("Sky News Australia", 28 Aug 2025)[14.3]. (Quote) "(...) The victims were aged between eight and ten".

14.4) *"Teen planning mass shooting arrested after weapons and manifesto found"* ("Daily Mail", 13 Sept 205)[14.4]. (Quote) "A 13-year-old was arrested after police found 23 guns, explosives, and a mass shooting manifesto. (...) The teen idolised school shooters and made threats to bomb or injure a school."

15) Mental health

15.1) *"Our children face dangerous double standard on behaviour, say educators"* ("WA Today", 27 Mar 2024)[15.1]. There is alarming data about students' plummeting mental health, their behaviour being among the worst in the world and vast numbers of teachers feeling unsafe. At schools, (Quote) "Violence and aggression are continually being played out." (end of quote). Numerous incidents of violence and aggression have been reported in classrooms, with students imitating the behaviours of their favourite actors, video game characters, and social media personalities.

16) Pressure to date and have sex

16.1) *"I didn't say no, but I regret it"* ("BBC", 6 Oct 2017)[16.1]. (Quote) "A 24-year-old secondary school teacher told the "BBC" she was shocked by the stories she heard from teenage pupils about their sexual activity". Recently, a friend of mine told me that her 5th-grade son receives bold love letters pressuring him to date these schoolgirls against his will.

17) School children identify themselves as animals

17.1) "*School children identifying as animals a 'societal problem'*" ("YouTube", "Sky News Australia", 6 July 2023)[17.1].
(Quote) "School head teacher C...B...S... discusses the controversy over children identifying as animals in schools and the need for guidance and moral leadership".

17.2) "*Parents, teachers warned to respect schoolkids Furries*" ("News.com.au", 26 June 2023)[17.2]. (Quote) "Teachers and parents are being warned not to ridicule or overreact to schoolkids identifying as animals".

18) Students' behaviour and other issues

I want to mention below a few stories published by "Sky News Australia" on traditional school topics:

18.1) "*Students' behaviour driving teachers away*" (Sept 2023)[18.1]

18.2) "*Declined significantly: Australia's education system has very large problems*" (Nov 2023)[18.2]

18.3) "*Overcrowded: Australian curriculum doesn't focus on essential knowledge, skills*" (Nov 2023)[18.3]

18.4) "*The kids are not alright: Murray discusses youth mental health crisis*" (Nov 2023)[18.4]

18.5) "*Primary school children feel more anxious than five years ago*" (Nov 2023)[18.5]

18.6) "*New extremes: viral videos reveal deviant schoolyard fight clubs*" (Oct 2023)[18.6]

18.7) "*Education system has 'big problem' with lack of discipline in classrooms*" (Nov 2023)[18.7]

18.8) "*Children need 'unstructured play' as a critical part of their human development*" (Feb 2023)[18.8]

19) News stories published by other Publishers:

19.1) "*It's time to teach Australian children how to behave, report says*" ("Sydney Morning Herald", Dec 2023)[19.1]

19.2) "*Predatory shops are selling lollies and vapes to children across Australia, with no date set for import ban*" ("The Guardian", Oct 2023)[19.2]

19.3) "*Vape stores clustered around schools and in the most disadvantaged suburbs, Australian study finds*" ("The Guardian", Jan 2024)[19.3]

19.4) "*Fear, shame, anger: child abuse inquiry hears of agony*" ("The Canberra Times", by William Ton, Oct 2023)[19.4]

19.5) "*Dad speaks out after daughter 12 dies following bullying*" ("The Sydney Morning Herald", 23 Sept 2024)[19.5]

19.6) "*Outrage over 'caged' autistic children*" ("SBS Australia", 28 Oct 2024)[19.6]. It reveals a shocking fact that some kids, especially those with special needs, were kept in literal cages. As a result of this form of exclusion, some children got injured.

19.7) "*Education Department to investigate allegations children were locked in cages at B...school*" ("ABC News", 22 Sept 2015)[19.7]

19.8) "*Families forced to remove children from school due to bullying*" ("ABC News" (AU), 25 Sept 2024)[19.8]

Articles like the above fuel my desire to homeschool our child and reinforce my parental determination to continue this path for as long as possible. I want to make decisions based on what is best for our child's mind, heart, and overall well-being, with the long term in mind.

Due to numerous problems within the traditional mainstream schooling system, many families have found a solution. According to the article in "The Educator" magazine, "Rise in homeschooling registrations a wake-up call for the system"[3.20], all states in Australia have experienced a noticeable increase in the number of children leaving the traditional school system in recent years. According to Dr David Roy at the University of Newcastle, (Quote) " (...) Many families have discovered through remote learning that their children not only survived but thrived in the home learning environment. Apart from philosophical reasons, special needs are the second most common reason for homeschooling (...). Research indicates there is no disadvantage to homeschooling in outcomes because it allows students to learn at their own pace and still excel, creating an immersive and engaging education environment" (end of quote). An example of a mother who, despite much opposition from her family and her own doubts, continued to homeschool her daughter successfully offers optimism and encouragement to those who have second thoughts about homeschooling. "I wanted to homeschool my daughter, but everyone in my family told me it was a bad idea. I did it anyway" ("<u>Insider</u>", 28/10/2025)[3.21] is a story of a parent who wanted her child to be safe from school dangers and have a more holistic education. It turned out to be a successful choice for her and her child.   *Please visit the "Copyright" page.

Tarasyasinski "Pixabay"

| 4 |

Find inspiration and motivation to homeschool

"Why not choose the less popular but proven way of education? You will never know its merit, unless you ever undertake it" (By Angel Claudius)

Tarasyasinski "Pixabay"

P arents and caregivers should have a strong motivation or reason behind their decision to home educate, which could make a huge positive difference in their endeavour. Motivation is essential for achieving the goal of successful home education and improving the homeschooling experience. It can be likened to the wind that propels a sailboat towards its destination. The stronger the motivation, the more likely the homeschooling experience will be uninterrupted and thriving.

Tarasyasinski "Pixabay"

Teaching at home may require some hard work, especially when starting up, and if you're not a trained teacher, you may also have to make a few sacrifices. However, many parents who have done it say it's worth it, and it even gets easier over time. This chapter outlines strategies to simplify the process of preparing for home education. The truth is that every child is unique, and each family has unique circumstances. So, take the time to evaluate your situation and determine whether home education is the right choice for you and your family. However, if it seems that is not the case, could you make adjustments to better fit your needs in this venture? Also, supporting children's education financially, whether in private or public schools, can be challenging for some families. For those who homeschool, a good option is for at least one parent to commit to being the primary caregiver and tutor. Arranging work schedules to accommodate home education is also feasible. I've heard of families successfully managing homeschooling while both parents worked full-time. One parent said she and her husband work on opposite shifts, so when one of them is at work, the other stays with the kids and homeschools them. In some countries, homeschooling parents may be eligible for financial assistance; therefore, it is advisable to check with the relevant government agency to see if this is available where you live. To make an informed decision to homeschool, please utilise the provided samples and tables below to consider all home education-related advantages and disadvantages. As the old saying goes: *"A goal without a plan is just a wish"* (Quote: Antoine de Saint-Exupéry).

The Pros and the Cons of Home Educating Your Child (Sample)

Advantages (Pros):

---

1. Provide a secure, carefully monitored learning environment and promote genuine, real-world learning experiences.

2. Play an essential role in safeguarding our children's identities.
3. Value spending quality time with our children to establish a family-oriented lifestyle that strengthens family bonds.
4. Prioritise and encourage the development of our children's passions and talents.
5. Create a calm and stress-free environment to help minimise anxiety and other intense emotions.
6. Foster a love for learning by giving students our full attention and providing one-on-one instruction.
7. Allow younger children ample playtime, and older children to pursue their personal interests and hobbies.
8. Facilitate learning at kids' own pace and in their preferred style and setting, tailored to their capabilities, special needs and personality.
9. Refrain from comparing our children to other students.
10. Minimise academic stress on our children.
11. Eliminate the chances and possibilities of our kids being bullied by peers and, at times, even teachers.
12. Avoid the grave dangers associated with attending mainstream schools, such as school violence, school shootings, bullying, stabbing and others.
13. Adhere to our preferred educational philosophy and learning styles.
14. Emphasise teaching our family's moral and spiritual values to children at home.
15. Provide personalised tutoring, instruction, and explanations by dedicated, intimately involved parents.

16. Save money on school fees and essential school resources, including textbooks, books, and formal uniforms. (Younger students can reuse the resources.)
17. Customise education to meet the children's educational, physical and emotional needs at their current level.
18. Provide a safe environment where learning is enjoyable and without boundaries.
19. Guide our children to help them choose the best association and make healthy friendships.
20. Involve our children in daily tasks and teach them the importance of responsibility throughout the day.
21. Offer our children more leisure time to explore their interests and hobbies so that they can experience a traditional childhood to the fullest.
22. Encourage natural and spontaneous learning.
23. Prioritise good emotional and physical health over academics.
24. Facilitate learning about the family's social traditions, spiritual inheritance, moral values and cultural customs.
25. Protect our children from medicinal drug abuse related to ADHD* or others. (*See the book by Ann Block, "*No More ADHD*")[4.1].
26. Protect our children against the possibility of incidents of sexual abuse at schools*. (*See story by William Ton in "Daily Mail", 22 October 2023)[4.2].
27. Protect our children from identity confusion. (In some schools, children are told they may believe they can become whatever they want to be, even animals. See chapter 3, point 17.10)[4.3]

28. Financial savings, plus other benefits that are priceless and cannot be acquired by finance...

The Pros and Cons of Home Educating Your Child (Sample)
Disadvantages (Cons):

---

1. Children may not easily understand or appreciate their parents taking on the role of their teachers.
2. Some worry that homeschooled children may receive a lower-quality education than children attending traditional, mainstream schools.
3. Only children may experience feelings of isolation and lack sufficient social interaction with peers (if allowed).
4. Balancing work and homeschooling can be challenging and even burdensome for some parents, who may need additional support as a result.
5. Creating a Homeschool Educational Plan or Curriculum can be a daunting and time-consuming activity for some parents.
6. Some children may lose interest and lack motivation in learning while being educated at home.
7. Only parents or caregivers may have limited time for secular work and other responsibilities.
8. Some parents may feel insecure and inadequate about their ability to teach their children effectively at home.
9. Some parents worry about their children's failure to meet the educational standards and goals typically expected of traditionally educated students.

10. Some to-be home educators are concerned about losing their professional identity and struggle to balance their homeschooling with secular careers and personal lives.
11. Homeschooling and parenting can be challenging, ongoing responsibilities.
12. Transition from homeschooling to mainstream school may be a challenge for some children.

> Sample motivation - Our Motivation to Home Educate is:

"We want to provide our child (children) with a stress-free learning environment to ensure they have the best possible start in life. We want to protect them from potential dangers in a traditional school setting or during their commute to and from school. It is important to us that our child fully embraces our family's cultural, spiritual and moral values (and others). Above all, we want to safeguard our child's heart, mind and identity. Spending more quality time with our child throughout the day, week, and year will strengthen our family bond, which is of primary importance to us".

Now, think about the Pros and Cons of homeschooling in your family's situation. Please list possible reasons and then categorise them into two columns. Then, create a "Solution Plan" to support your child's home education by outlining solutions for each problem or disadvantage identified.

Quote: "No amount of money or success can replace time spent with your family" (Author unknown)

> An example of a Solution:

Disadvantage: When parents take on the role of a teacher for their children, they (the students) may feel uncomfortable with this arrangement.

Solution: Explain that the parent will serve as the home educator. This can be beneficial for many reasons. There will be no bullying, the education can be made fun and tailored to the child's needs and personal interests. Learning in a home environment will allow the child to spend more time with family and friends, have more time for play and travel, and engage in other fun and engaging activities. Other homeschooling families could encourage the child and demonstrate that the homeschooling approach has been successful for them.

Tarasyasinski "Pixabay"

| 5 |

Conclusion

"Cherish each day with your precious children while homeschooling, as they will surely be gone before you know it" (By Angel Claudius)

Franz26 "Pixabay"

Good advice on successful homeschooling is abundant. While I genuinely enjoy homeschooling, there are certain aspects of it that I don't find as enjoyable. However, like with any work, it's impossible to love everything about it. Just as it is with a slow-growing tree, we must cultivate patience to see it grow. In a similar way, home educators must cultivate love and patience, knowing that, with the kids around all the time, they will frequently disrupt their peace. But remember, they are more important than your to-do list or anything else!

Tarasyasinski "Pixabay"

As a parent, you want the best for your children and want them to enjoy a happy, peaceful and rewarding life, not only during childhood but also in adulthood. Homeschooling will bring you closer to that goal and even make this possible. The homeschooling education is becoming more popular these days since the violence and crime at schools are on the increase. We see a great deal of news about adverse events in mainstream schools worldwide. I felt very sad after learning that one mother's child had not only been physically abused at high school, but was also fearful for his life after receiving death threats from students of that same school. Recently, there have been a few teen deaths by stabbing in the community, and the family's fear for the child's safety was very real.

Parents lack control over their children's education at schools and cannot withdraw them from subjects they find unpopular, sensitive, or even contrary to their beliefs. Not only that, the stress and anxiety rates of primary and high school children attending mainstream schools have been rising drastically, to the extent that some even refuse to go to school! Violence in many schools is becoming an everyday thing, and parents worry about what is going to happen to their child during their absence. School associations are shaping children to become very unkind, disobedient, sometimes even violent, use foul language, and practice sex before marriage in their teen years, among other things. Sadly, school years have the great potential to erode the good morals, values, and habits that many families try so hard to instil in their children at home. Moreover, children with disabilities feel at times excluded from class activities and are bullied, and even held in protective cages in some instances (where they usually get physically and emotionally hurt). Because of the use of social media, many kids at school (and outside) are cyberbullied, which takes away children's happiness and negatively impacts their mental health for the rest of their lives. In a few cases, some are stabbed with sharp objects by so-called school friends. And the list goes on...

The story titled " 'Adolescence' writer reveals school girl's sad reaction" ("Metro", 22/09/2025)[5.1], relates to the drama recently released on "Netflix". It's about a 13-year-old boy accused of murdering a girl from his school. The drama (Quote) "sparked conversations about the 'manosphere', misogyny and violence against women and girls and the widespread use of social media among younger generations" (end of quote). The co-writer and actor of the drama (Quote) "had the idea for the story after hearing about real incidents of knife crime committed by young boys against young girls" (end of quote).

Home education eliminates most of the existing problems with behaviour and education in mainstream schools. When homeschooling for most of the day, these kids can utilise practical life experiences, learning materials at their level and in their preferred way and style, at their own pace. Most of all, parents have complete control over the child's development and education, can teach them the family's moral values and cultural heritage, and model good behaviour. Parents can shape their children's childhoods into happy, traditional ones.

TyliJura "Pixabay"

Throughout this publication, I've shared tips based on my own experience and others', backed by research and media articles. These tips and information may be practical for specific individuals and touch on the surface of available options. Typically, each home educator learns through trial and error, but it doesn't have to be that way. My motto has always been "*Learn from someone else's mistakes, not your own".* I hope that by sharing the valuable lessons throughout this publication, I have inspired you and others to succeed as home educators.

In this book, I discuss the everyday challenges faced by mainstream schooling, whether public or private. My main goal and intention have not been to criticise any particular school, institution, organisation, or education system, no!, but to demonstrate the superiority of home education as an educational model and to provide the facts. Even with reforms to the curriculum or school safety and others at mainstream schools, these changes alone are insufficient to address the various other issues that are, in fact, increasing ("Sweeping reforms to hit Aussie public schools", "News Wire", 18 October 2025[5.2]; "Sweeping reforms set to hit every school across Australia", "Journo", 20 October 2025)[5.3].

So, friends, embrace your new journey as a homeschooling family and enjoy the blessing of bonding with your children through the fantastic home education lifestyle. Please always remember that there's no job more important than raising and teaching your children the right, holistic way—at home. Always keep in mind the long view for your kids' home education to succeed. Most of all, cherish each day with your precious children during your homeschooling adventure, because *"they will be gone before you know it"*!

Angel

| 6 |

Resources: Useful Sites

"We reach out to resources for quick and effective solutions" (By Angel Claudius)

TyliJura "Pixabay

Online Live and Pre-recorded Classes:

1. https://www.blindmewithscience101.com - Live science lessons for kids of all ages via Zoom with Paul
2. https://www.drewsalemhistory.com - Live history lessons for kids of all ages and adults (covering international history) via Zoom with Drew
3. Email: teachercindyatgrasp@gmail.com - English and Drama live lessons via Zoom with Cindy
4. https://www.afroginoz.com.au - French live lessons via Zoom with Sandrine

Useful Sites:

1. https://www.citepublisher.com - educational publications for children and adults, especially for managing emotions in children and homeschooling
2. https://www.angelclaudius.com - homeschooling and educational publications beneficial in home education
3. https://hslda.org - international website supporting homeschooling
4. https://www.spelfabet.com - 'Spelfabet', learning the building blocks of word-sounds, their spelling and word parts
5. https://www.usborne.com/quicklinks - Usborne books and links to internet Sites
6. https://www.education.com
7. https://www.montessoriprintshop.com

8. https://www.au.ixl.com - "IXL" Personalised Maths and English learning
9. https://www.homeschoolmath.net
10. https://www.mathsonline.com.au
11. https://www.teachersuperstore.com.au
12. https://www.familyeducation.com/school-learning/learning-styles - Learning styles
13. https://artventure.com.au
14. https://howtohomeschoolforfree.com/full-online-homeschool-curriculum - Free homeschool online curriculum
15. https://readyed.net/product-category/online-classroom/ - Online Classroom: Australian history, geography, mammals, bullying
16. https://www.kiwico.com - science, art, engineering
17. https://www.crayola.com - free art lesson plans
18. https://www.juniorengineers.com.au - coding and robotics
19. https://www.fivesenseseducation.com.au - homeschooling resources supporting the Australian curriculum

Languages:

1. https://dinolingo.com - Dinolingo
2. https://duolingo.com - Duolingo
3. https://ling-app.com - Ling
4. https://www.innovativelanguage.com/homeschool

WHY CHOOSE HOMESCHOOLING? - 49

International Online Schools:

1. https://brisbanesde.eq.edu.au
2. https://www.tekura.school.nz
3. https://cairnssde.eq.edu.au - Global Learning
4. https://www.khanacademy.org - free school
5. https://www.teaching.com.au
6. https://www.lessonzone.com.au
7. https://www.ck12.org
8. https://www.abc.net.au - ABC Education
9. https://www.remotelearning.school
10. https://wideopenschool.org

Australian Support Services:

1. https://www.hea.edu.au - "The Home Education Association" (HEA) - Provides legal advice and a free hotline. It helps homeschoolers across Australia
2. https://www.hea.edu.au/free - Free homeschooling resources
3. https://home-ed.vic.edu.au - "The Home Education Network Victoria"
4. https://hewa.wa.edu.au - "The Home Education WA"
5. https://shen.org.au - "The Sydney Home Education Network" (SHEN)
6. https://www.yeahs.vic.edu.au/ayce - "Yea High School's AYCE Program" supporting homeschoolers at high school level in Victoria State, Australia

Homeschooling publications

If you would like to find out HOW to homeschool, please explore the publications below from the homeschooling series available from https://www.angelclaudius.com

1. **ISBN 978-0-6458672-5-1** "Homeschool Easy: A Guide to Inspire and Support Homeschoolers and Home Educators with Proven Fifty Tips and Helpful Resources".
2. **ISBN 978-0-6455541-1-3** "Homeschooling in a Nutshell: Effective Advice for Busy Parents to Achieve Success in Home Education".
3. **ISBN 978-0-6455541-9-9** "The Successful Home Education Guide: Inspiring and Supporting Home-Based Schooling with Proven Tips and Practical Printable Resources".
4. **ISBN 978-0-6458672-0-6** "Writing a Homeschool Curriculum: A Step-By-Step Guide to Writing a Homeschool Learning Plan with Essential Resources".
5. **ISBN 978-0-6458672-1-3** "Homeschool Curriculum Journal: Essential Record Pages for Writing a Detailed Homeschool Learning Plan".
6. **ISBN 978-0-6458672-6-8** "Motivating a Homeschooler to Study, How?: Helping Students Enjoy Learning at Their Home-Based Schools".

7. **ISBN 978-0-6458672-8-2** "Succeeding in Teaching at Home-Based Schools: Simple and Effective Teaching Methods and Techniques for Home Educators".

Homeschooling publications for children, available from https://www.angelclaudius.com

1. **ISBN 978-0-6458672-2-0** "Why I Love Homeschooling: Why Homeschooled Kids Are Happy Learners in Home-Based Schools".
2. **ISBN 978-0-6458672-3-7** "Why I Love Homeschooling: Why Homeschooled Kids Are Happy Learners in Home-Based Schools; Handwriting Practice Workbook". - A handwriting workbook for homeschooled children to learn and practice handwriting.
3. **ISBN 978-0-6458672-7-5** "Why I Stay Motivated to Study and Learn in My Home-Based School: How motivation and enjoyment of life help young learners in homeschooling".

Other homeschooling publications are available from both www.citepublisher.com and www.amazon.com

1. **ISBN 9780645554137** (Paperback), "My Daily Checklist: The homeschooler's record-keeping diary";
1. **ISBN 9780645180602** (Paperback), "I can surf the waves of strong emotions: A parent-child guide to help older children cope with strong emotions";

2. **ISBN 97880645180619** (Hardback), "I can surf the waves of strong emotions: A parent-child guide to help older children cope with strong emotions";
3. **ISBN 9780645180688** (Paperback), "I can surf the waves of strong emotions: The Montessori-inspired handwriting practice workbook to help children handle strong emotions";
4. **ISBN 9780645180626** (Paperback), "I can surf the waves of strong emotions: A companion workbook to help older children take control of strong emotions";
5. **ISBN 9780645180640** (Paperback), "I can surf the waves of strong emotions: Affirmation posters to help older children cope with strong emotions";
6. **ISBN 9780645554106** (Paperback), "Creative letter-writing: The Montessori-inspired colour-in letterheads for children's letter-writing in cursive handwriting";

Bible Education

1. **ISBN 9780645180657** (Paperback), "The Bible stories for preschoolers: The Montessori-inspired handwriting & colour-in activity workbook, to help small children become familiar with the Bible";
2. **ISBN 9780645554151** (Paperback), "Learning the lessons from Bible stories: The Montessori-inspired handwriting & colour-in activity workbook, part 1". This publication is for elementary-level children.

References

Chapter 1

1.1) https://www.youtube.com/watch?v=Ei_v08ywDGg

1.2) https://www.msn.com/en-au/news/other/i-homeschool-my-kids-parents-like-me-are-judged-and-exhausted/ss-AA1LnRWN?ocid=msedgntp&pc=DCTS&cvid=ce5f1ef3e2774b76a27b4812eafe815b&ei=44#image=9

Chapter 2

2.1) https://www.msn.com/en-au/health/other/experts-reveal-side-effects-of-prescribing-adhd-medication-too-early-in-a-child-s-life/ar-AA1LvwiZ?ocid=msedgntp&pc=DCTS&cvid=ce5f1ef3e2774b76a27b4812eafe815b&ei=58

2.2) https://www.nbcnews.com/now/video/-unschooling-movement-has-parents-teaching-kids-outside-of-classrooms-245376069775

2.3) https://shows.acast.com/rosebud-with-gyles-brandreth-new-episodes/anthony-horowitz

2.4) https://www.businessinsider.com/friends-family-told-us-not-homeschool-kids-we-didnt-listen-2025-5

2.5) https://www.smh.com.au/national/victoria/i-can-t-learn-why-some-parents-have-swapped-school-for-homeschooling-20240712-p5jt70.html

2.6) https://asoftgentlevoice.blogspot.com/2008/07/meat-for-mind.html

Chapter 3

1.1) https://canberratimes.com.au/story/8395923/fear-shame-anger-child-abuse-inquiry-hears-of-agony/

2.1) https://www.abc.net.au/news/2025-01-15/australian-families-struggling-to-afford-their-kids-education/104814986

54 - REFERENCES

https://www.adogs.info/press/how-free-are-public-schools. How free are public schools? Australian Council for the Defence of Government Schools.

2.2)https://www.reddit.com/r/australia/comments/1kb4963/public_school_parents_are_you_aware_of_the/?rdt=55511

2.3)https://www.theeducatoronline.com/k12/news/costofliving-squeeze-hits-school-families/286782

3.1)https://www.skynews.com.au/opinion/school-children-asked-to-complete-outrageous-survey-on-their-gender-and-pronouns/video/f136af88999f960b344270ae8777

3.2)https://www.skynews.com.au/opinion/chris-kenny/clutching-at-any-diagnosis-parents-look-for-help-with-childrens-mental-health/video/8cd0b69f1cb429f8b41b817fa2984b11

3.3)https://www.skynews.com.au/opinion/rita-panahi/scary-stuff-hospitals-send-genderconfused-kids-to-undergo-transiton/video/cd51f72336f8b95182acdd9c7faddd60

3.4)https://www.skynews.com.au/opinion/virtue-signalling-highschool-converts-femaleonly-toilet-block-to-unisex/video/89a7614314396bca24f4544715652614

4.1)https://www.skynews.com.au/australia-news/victorian-schools-facing-mass-teacher-shortages/video/c9ffe9dd8dbbec691497aa72f854c079

4.2)https://www.skynews.com.au/politics/australi-in-the-middle-of-an-education-staffing-crisis/video/3e51f0725da51019761c0d996d9dee39

4.3)https://www.theguardian.com/australia-news/2025/aug/06/qld-teachers-strike-qtu-union-queensland

5.1)https://www.skynews.com.au/opinion/andrew-bolt/walking-billboard-victorian-schools-new-uniform-has-a-political-agenda/video/437ed2245d97ac9c32b2da77e54adbcb

5.2)https://www.skynews.com.au/opinion/paul-murray/parents-are-desperately-concerned-about-ideology-taught-in-schools/video/233e7f30a19fa7fac35308807c00bdce

6.1)https://www.skynews.com.au/opinion/shouldnt-be-surprising-the-school-system-is-broken-caleb-bond/video/dfc7d65c7fc1d842b1e3c51920513eb7

6.2)https://www.abc.net.au/news/2023-08-23/one-in-three-students-not-meeting-naplan-standards/102756262

6.3) https://www.skynews.com.au/opinion/paul-murray/misery-taught-in-schools-stopping-students-from-striving-to-achieve/video/32f0a275bfc3f68af1731b2346217506

7.1) https://www.smh.com.au/national/concern-after-ex-students-enter-school-with-weapons-20230812-p5dw0n.html

8.1) https://www.dailymotion.com/video/x8n87xp

8.2) https://www.skynews.com.au/australia-news/childrens-worries-could-turn-into-longterm-mental-health-conditions-if-not-managed-video/16b93cb5136b38434f90fa2e7f31257d

8.3) https://www.skynews.com.au/opinion/rita-panahi-growing-number-of-school-students-becoming-paralysed-by-social-anxiety/video/9172b9816efa48dbaec4a554a8c5cd27

8.4) https://www.mouthsofmums.com.au/blog

9.1) https://www.dailymotion.com/video/x8nbzzx

9.2) https://www.msn.com/en-au/news/australia/aussie-mum-makes-global-plea-at-un-to-kick-kids-off-social-media/vi-AA1NkT5V?ocid=msedgntp&pc=DCTS&cvid=1cdc3df2068d48b1fa718f7c18e9379b&ei=45

10.1) https://www.abc.net.au/news/2023-08-17/research-finds-disabled-students-endure-bullying-and-exclusion/102744586

10.2) https://www.abc.net.au/news/2024-11-20/bullied-children-moved-teachers-struggle-to-control-students/104589574

11.1) https://www.sbs.com.au/news/article/several-people-seriously-injured-after-school-bus-roll-over-in-regional-victoria-town/svrtdexub

11.2) https://www.bbc.com/news/articles/cqldr37n2xxo

11.3) https://www.msn.com/en-us/news/world/at-least-6-students-drown-24-others-injured-during-school-trip-to-a-beach/ar-AA1L5gRo

11.4) https://www.abc.net.au/news/2025-10-13/thom-hosking-remembered-as-glue-of-family-bendigo-crash/105885680

12.1) https://www.lbc.co.uk/article/protest-erupts-school-prison-rules-students-suspended-DWzgJw_2/

12.2) https://www.news.com.au/lifestyle/real-life/twin-tragedy-dads-grief-as-4yo-is-crushed-to-death-at-school-birthday-party/news-story/ae1ef37e7cef661edb6f9512d0cde9c7

12.3) https://www.abc.net.au/news/2025-09-01/auburn-south-primary-jack-davey-court-case/105721190

56 - REFERENCES

12.4) https://www.msn.com/en-au/news/other/parents-anger-at-new-prison-like-rules-at-school/ar-AA1M7eWd?ocid=msedgntp&pc=DCTS&cvid=68bfa61c39244f61b44d5cbb641d4d8f&ei=16

13.1) https://www.abc.net.au/news/2024-04-29/school-refusal-cant-australia-education-four-corners/103669970

13.2) https://www.skynews.com.au/lifestyle/health/school-refusal-closely-related-to-anxiety-in-young-children-and-preteens/video/0feeb12a5e0271a8b599c8eec1d747c2

14.1) https://abcnews.go.com/US/116-people-died-gun-violence-day-us-year/story?id=97382759

14.2) https://www.euronews.com/my-europe/2025/06/10/at-least-five-killed-in-school-shooting-in-austria-reports-say

14.3) https://www.abc.net.au/news/2025-08-29/parents-of-child-killed-in-minneapolis-shooting-want-gun-control/105711254

14.4) https://www.msn.com/en-au/news/other/teen-planning-mass-shooting-arrested-after-weapons-and-manifesto-found/vi-AA1Mdb82?ocid=msedgntp&pc=DCTS&cvid=68c5175900854c6691a06b3a7504b646&ei=18

11.5) https://www.abc.net.au/news/2025-11-14/islah-metcalfe-rouse-hill-public-school-two-more-students-struck/106007398

15.1) https://www.watoday.com.au/national/western-australia/our-children-face-dangerous-double-standard-on-behaviour-say-educators-20240325-p5ff51.html

16.1) https://www.bbc.com/news/education-41512300

17.1) https://www.skynews.com.au/opinion/piers-morgan/school-children-identifying-as-animals-a-societal-problem/video/16a8cf41aa466675f7532deb8bdf2428

17.2) https://www.news.com.au/lifestyle/parenting/school-life/parents-and-teachers-warned-not-to-ridicule-schoolkids-who-identify-as-animals/news-story/fea6c5db9da4854e5abe3782893dbb3d

18.1) https://www.skynews.com.au/opinion/students-behaviour-driving-teachers-away/video/1a2eb6e4093bf08136bae01c5c11437e

18.2) https://www.skynews.com.au/australia-news/declined-significantly-australias-education-system-has-very-large-problems/video/e15fc0dea3d287af5faa5d8c60db1e9b

REFERENCES - 57

18.3) https://www.skynews.com.au/opinion/peta-credlin/overcrowded-australian-curriculum-doesnt-focus-on-essential-knowledge-skills/video/3b9b8241c648fab558871b436730a3c4

18.4) https://www.skynews.com.au/opinion/paul-murray/the-kids-are-not-alright-murray-discusses-youth-mental-health-crisis/video/f1e5239ff95466f89720742a6d812d94

18.5) https://www.skynews.com.au/lifestyle/health/primary-school-children-feel-more-anxious-than-five-years-ago/video/605b8e75c01ee296c62a5e89921c63d8

18.6) https://www.skynews.com.au/opinion/chris-kenny/new-extremes-viral-videos-reveal-deviant-schoolyard-fight-clubs/video/c5b87f8ef37d7ca39308c109dbe8ecf9

18.7) https://www.skynews.com.au/opinion/education-system-has-big-problem-with-lack-of-discipline-in-classrooms/video/a2a2600ebd14c46808fdb180c5814748

18.8) https://www.skynews.com.au/opinion/chris-kenny/kids-need-unstructured-play-as-a-critical-part-of-their-human-development/video/a7185ceb64f320bea5126ca642dd63ba

19.1) https://www.smh.com.au/national/nsw/it-s-time-to-teach-australian-children-how-to-behave-report-says-20231201-p5eodb.html

19.2) https://www.theguardian.com/australia-news/2023/oct/02/predatory-shops-are-selling-lollies-and-vapes-to-children-across-australia-with-no-date-set-for-import-ban

19.3) https://www.theguardian.com/australia-news/2024/jan/31/vape-stores-concentrated-around-schools-and-socioeconomically-disadvantaged-suburbs-australian-study-finds

19.4) https://www.canberratimes.com.au/story/8395923/fear-shame-anger-child-abuse-inquiry-hears-of-agony

19.5) https://www.smh.com.au/national/dad-speaks-out-after-daughter-12-dies-following-bullying-20240923-p5kcvr.html

19.6) https://www.sbs.com.au/news/article/outrage-over-caged-autistic-children/pwzjpobv8

19.7) https://www.abc.net.au/news/2015-09-22/dept-investigating-claims-bendigo-students-were-locked-in-cages/6795966

19.8) https://www.dailymotion.com/video/x967qou

58 - REFERENCES

3.20)https://www.theeducatoronline.com/k12/news/rise-in-home-schooling-registrations-a-wakeup-call-for-the-system/279483 "Rise in homeschooling registrations a 'wake-up' call for system", "The Educator"

3.21)https://africa.businessinsider.com/news/i-wanted-to-homeschool-my-daughter-but-everyone-in-my-family-told-me-it-was-a-bad/gffb9v6

Chapter 4

4.1)https://www.independent.co.uk/bulletin/lifestyle/adhd-medication-prescribed-children-too-early-b2816803.html

4.2)https://www.canberratimes.com.au/story/8395923/fear-shame-anger-child-abuse-inquiry-hears-of-agony/

4.3)https://www.skynews.com.au/opinion/piers-morgan/school-children-identifying-as-animals-a-societal-problem/video/16a8cf41aa466675f7532deb8bdf2428

Chapter 5

5.1)https://www.msn.com/en-au/news/other/adolescence-writer-reveals-schoolgirl-s-sad-reaction/ar-AA1MWB8J?ocid=msedgntp&pc=DCTS&cvid=68d0f6da3a8e499c928eafb5d1b258b2&ei=35

5.2)https://www.msn.com/en-au/news/other/sweeping-reforms-to-hit-aussie-public-schools/ar-AA1OHaQm

5.3)https://www.kucera.news/2025/10/18/sweeping-reforms-set-to-hit-every-school-across-australia/

TyliJura "Pixabay"

www.ingramcontent.com/pod-product-compliance
Lightning Source LLC
LaVergne TN
LVHW031604060526
838200LV00055B/4484